Marine Renewable Energy Systems Engineering Solutions for a Sustainable Future

Gilbert

Copyright © [2023]

Title: Marine Renewable Energy Systems Engineering Solutions for a Sustainable Future
Author's: Gilbert

This book was printed and published by [Publisher's: **Gilbert**] in [2023]

ISBN:

TABLE OF CONTENT

Chapter 1: Introduction to Marine Renewable Energy Systems

Overview of Marine Renewable Energy

Importance of Sustainable Energy Solutions

Challenges and Opportunities in Marine Renewable Energy Systems

Chapter 2: Fundamentals of Coastal and Offshore Engineering

Coastal Processes and Erosion

Offshore Engineering Principles

Environmental Considerations in Coastal and Offshore Engineering

Chapter 3: Design and Management of Coastal Structures

Breakwaters and Seawalls

Revetments and Groynes

Beach Nourishment and Sediment Management

Chapter 4: Design and Management of Offshore Platforms 28

Fixed Offshore Platforms

Floating Offshore Platforms

Subsea Structures and Foundations

Chapter 5: Introduction to Marine Renewable Energy Systems 34

Overview of Renewable Energy Sources

Types of Marine Renewable Energy Systems

Advantages and Disadvantages of Marine Renewable Energy

Chapter 6: Wave Energy Conversion Systems 40

Wave Characteristics and Measurement

Wave Energy Conversion Technologies

Design and Optimization of Wave Energy Converters

Chapter 7: Tidal Energy Conversion Systems   46

Tidal Characteristics and Measurement

Tidal Energy Conversion Technologies

Design and Optimization of Tidal Energy Converters

Chapter 8: Ocean Current Energy Conversion Systems   53

Ocean Current Characteristics and Measurement

Ocean Current Energy Conversion Technologies

Design and Optimization of Ocean Current Energy Converters

Chapter 9: Offshore Wind Energy Systems   60

Wind Characteristics and Measurement

Offshore Wind Turbine Technologies

Design and Optimization of Offshore Wind Farms

Chapter 13: Conclusion 84

Summary of Key Findings

A Call for Collaboration and Sustainable Solutions in Marine Renewable Energy Systems

Chapter 1: Introduction to Marine Renewable Energy Systems

Overview of Marine Renewable Energy

Marine renewable energy refers to the harnessing of energy from the ocean, including waves, tides, and currents, to generate electricity. As the world seeks sustainable alternatives to fossil fuels, marine renewable energy has emerged as a promising solution, offering a vast and untapped source of clean, renewable power.

This chapter provides an overview of marine renewable energy, focusing on its engineering solutions and their relevance to the field of geological engineering. Geological engineers play a crucial role in assessing the geological characteristics of potential marine energy sites, ensuring their stability and viability for energy extraction.

One of the key forms of marine renewable energy is wave energy, which involves capturing the energy generated by ocean waves. Engineers have developed various technologies to harness this power, including wave energy converters that convert the kinetic energy of waves into electricity. These converters can be deployed near coastlines or offshore, making wave energy a versatile and abundant resource for coastal regions.

Tidal energy is another important component of marine renewable energy. Tidal currents are a predictable and consistent source of energy, making them highly attractive for power generation. Engineers have designed tidal turbines to extract energy from tidal currents, similar to wind turbines but placed underwater. These turbines require in-depth geological assessments to ensure their proper installation and operation, making the expertise of geological engineers invaluable.

In addition to waves and tides, marine currents also offer significant potential for energy extraction. Strong ocean currents, such as those found in narrow straits or channels, can be harnessed using underwater turbines. Geological engineers play a critical role in mapping these currents and identifying suitable locations for turbine deployment.

Furthermore, this chapter delves into the challenges and opportunities associated with marine renewable energy. Engineering solutions to mitigate the potential environmental impacts of marine energy systems are discussed, including the need for comprehensive environmental impact assessments and monitoring programs.

Overall, this subchapter provides a comprehensive introduction to marine renewable energy, highlighting its relevance to geological engineering. By leveraging the engineering expertise of geological engineers, the sustainable development of marine renewable energy systems can be achieved, paving the way for a cleaner and more sustainable future.

Importance of Sustainable Energy Solutions

As engineers specializing in geological engineering, it is crucial to understand the significance of sustainable energy solutions in today's world. The ongoing global energy crisis and the environmental challenges associated with traditional energy sources have made it imperative for engineers to explore alternative and sustainable solutions. This subchapter aims to shed light on the importance of sustainable energy solutions and their relevance to geological engineering.

Sustainable energy solutions, such as marine renewable energy systems, offer numerous benefits that make them essential for engineers in the field of geological engineering. Firstly, these solutions help reduce the reliance on fossil fuels, which are finite resources and contribute significantly to climate change. By harnessing the power of marine resources, such as tides, waves, and offshore wind, engineers can ensure a continuous and abundant supply of clean energy without depleting the Earth's resources.

Furthermore, sustainable energy solutions provide a way to mitigate the environmental impact of traditional energy sources. Geological engineering often involves working in delicate ecosystems, and the use of fossil fuels can have detrimental effects on these environments. By implementing marine renewable energy systems, engineers can significantly reduce greenhouse gas emissions, minimize air and water pollution, and protect biodiversity.

Another crucial aspect of sustainable energy solutions is their long-term economic viability. Investing in renewable energy technologies creates job opportunities and stimulates economic growth. As engineers, it is essential to recognize the potential for innovation and technological advancements in the field of marine renewable energy systems. By driving research and development in this area, engineers can contribute to the growth of a sustainable economy.

Moreover, sustainable energy solutions enhance energy security and resilience. Geological engineers often work in remote locations or areas prone to natural disasters. By utilizing renewable energy sources, such as offshore wind farms or wave energy converters, engineers can ensure a reliable and independent source of power, reducing the vulnerability of communities and infrastructure to energy supply disruptions.

In conclusion, the importance of sustainable energy solutions cannot be overstated for engineers specializing in geological engineering. By adopting marine renewable energy systems, engineers can contribute to the transition towards a more sustainable future. These solutions offer environmental benefits, economic opportunities, and increased energy security, making them vital tools for addressing the global energy crisis and mitigating the impact of traditional energy sources. As engineers, it is our responsibility to embrace and promote sustainable energy solutions to ensure a sustainable and resilient future for generations to come.

Challenges and Opportunities in Marine Renewable Energy Systems

As engineers in the field of geological engineering, we are at the forefront of developing sustainable solutions for the world's growing energy needs. One of the most promising avenues for achieving this goal lies in marine renewable energy systems. Harnessing the power of the ocean holds immense potential, but it also presents unique challenges that require innovative engineering solutions.

One of the primary challenges in marine renewable energy systems is the harsh and unpredictable nature of the marine environment. The constant exposure to waves, tides, and currents can pose significant structural and operational risks. Engineers must design robust systems capable of withstanding extreme weather conditions, corrosion, and erosion. This calls for the use of advanced materials, such as corrosion-resistant alloys and protective coatings, as well as robust monitoring and maintenance strategies.

Another key challenge is the efficient conversion of marine energy into usable electricity. Different technologies, such as tidal turbines, wave energy converters, and offshore wind farms, each have their own technical complexities. Engineers must optimize the design, placement, and operation of these systems to maximize power generation while minimizing environmental impacts. This requires a deep understanding of fluid dynamics, structural mechanics, and electrical engineering principles.

Furthermore, the deployment and maintenance of marine renewable energy systems present logistical and cost challenges. Installing and maintaining offshore structures can be expensive and technically demanding. Engineers must develop innovative installation methods, such as floating platforms or modular designs, to reduce costs and increase efficiency. Additionally, ensuring the long-term reliability and performance of these systems requires careful planning and ongoing monitoring.

Despite these challenges, there are exciting opportunities for engineers in the field of marine renewable energy systems. As the demand for clean and sustainable energy continues to grow, there is a significant need for engineers with expertise in this field. This presents a unique opportunity for geological engineers to apply their knowledge of the marine environment and geological processes to develop innovative and efficient energy solutions.

Moreover, the development of marine renewable energy systems can have positive environmental and socio-economic impacts. By reducing reliance on fossil fuels, these systems contribute to mitigating climate change and improving air quality. Additionally, the deployment of such systems can create new job opportunities, stimulate local economies, and enhance energy security.

In conclusion, the challenges and opportunities in marine renewable energy systems are vast and complex, requiring the expertise of engineers specializing in geological engineering. By developing innovative solutions to overcome the challenges posed by the marine environment, engineers can contribute to a sustainable future while reaping the benefits of a rapidly growing field.

Chapter 2: Fundamentals of Coastal and Offshore Engineering

Coastal Processes and Erosion

Introduction to Coastal Processes and Erosion

Understanding coastal processes and erosion is crucial for engineers specializing in geological engineering. Coastal areas are dynamic environments where land and sea interact, resulting in various processes that shape the coastline. This subchapter aims to provide engineers with an overview of these processes and the challenges they pose to coastal structures. By understanding these dynamics, engineers can develop sustainable solutions for coastal regions and contribute to the field of marine renewable energy systems.

Coastal Processes

Coastal processes encompass a range of physical, chemical, and biological phenomena that occur at the interface of land and sea. These processes include wave action, tides, currents, sediment transport, and weathering. Engineers must understand how these processes interact and affect the coastal environment to design effective structures and systems.

Wave Action and Erosion

Waves are a dominant force in coastal regions and play a significant role in erosion. Engineers must analyze wave characteristics, such as height, period, and direction, to assess their erosive potential. Understanding wave energy patterns helps in determining suitable locations for marine renewable energy systems, such as wave energy converters.

Tides and Currents

Tides and currents also influence coastal erosion. Engineers need to study tidal patterns and understand how they affect sediment transport and erosion rates. This knowledge is crucial when designing coastal structures like breakwaters or determining the optimal location for tidal energy devices.

Sediment Transport and Deposition

Sediment transport is a complex process that engineers must consider when designing coastal structures. Understanding the movement of sediment, including sand, gravel, and silt, is essential for predicting erosion and deposition patterns. Engineers must assess the sediment budget and develop strategies to mitigate erosion and maintain a stable coastline.

Coastal Erosion and Engineering Solutions

Coastal erosion poses significant challenges for engineers in designing and maintaining coastal infrastructure. The erosive forces acting on coastlines can threaten properties, ecosystems, and human lives. Engineers specializing in geological engineering play a crucial role in developing sustainable solutions to combat erosion.

Coastal Protection Measures

Various strategies can be employed to protect coastal areas from erosion. These include the construction of seawalls, groins, breakwaters, and beach nourishment. Engineers must assess the effectiveness, cost, and environmental impact of these measures to select the most suitable solution for a given coastal area.

Conclusion

Coastal processes and erosion are vital considerations for engineers specializing in geological engineering. Understanding the dynamic interactions between land and sea allows engineers to design resilient coastal structures and contribute to the development of sustainable marine renewable energy systems. By addressing the challenges of coastal erosion, engineers can help protect the environment, preserve coastal communities, and ensure a sustainable future for coastal regions.

Offshore Engineering Principles

In the rapidly evolving field of marine renewable energy systems, offshore engineering principles play a crucial role in ensuring the successful implementation and operation of sustainable solutions. This subchapter aims to provide engineers, particularly those specializing in geological engineering, with a comprehensive understanding of the fundamental principles that govern offshore engineering.

Geological engineering, being a niche within engineering, focuses on understanding the geological aspects of the earth's surface and how they interact with engineering structures. In the context of offshore engineering, this field becomes even more significant due to the unique challenges posed by the marine environment.

The subchapter begins by introducing the basic concepts of offshore engineering, emphasizing the importance of site investigation and assessment. Engineers must have a thorough understanding of the geological conditions, such as seabed composition, sediment dynamics, and geohazards, to design and construct offshore structures that can withstand the harsh marine environment.

The subchapter then delves into the various types of offshore structures commonly used in marine renewable energy systems, such as offshore wind turbines, wave energy converters, and tidal energy devices. For each type, the geological considerations and engineering principles specific to their design and installation are discussed. This includes aspects like foundation design, anchoring systems, and the impact of wave and current forces on the structure.

Furthermore, the subchapter explores the principles of offshore geotechnical engineering, which involve analyzing the behavior of soils and rocks under marine conditions. Topics covered include soil mechanics, seabed stability, and the selection of appropriate foundation systems based on geological characteristics.

To ensure the safe and efficient operation of marine renewable energy systems, the subchapter also addresses the principles of offshore monitoring and maintenance. This includes the use of remote sensing technologies, underwater robotics, and advanced data analysis techniques to assess the structural integrity and performance of offshore structures.

Throughout the subchapter, real-life case studies and examples from offshore projects around the world are incorporated to provide practical insights and enhance the understanding of the engineering principles discussed.

By the end of this subchapter, engineers specializing in geological engineering will have gained a comprehensive understanding of the offshore engineering principles necessary for designing, constructing, and maintaining marine renewable energy systems. With this knowledge, they will be better equipped to contribute to the development of sustainable solutions that harness the power of the ocean while minimizing environmental impact.

Environmental Considerations in Coastal and Offshore Engineering

As engineers in the field of geological engineering, it is crucial to recognize and address the environmental considerations in coastal and offshore engineering. The delicate balance between harnessing renewable energy and preserving our natural ecosystems requires careful planning, advanced technologies, and a deep understanding of the environmental impact.

Coastal and offshore engineering projects, such as marine renewable energy systems, play a vital role in our transition towards a sustainable future. However, these projects must be undertaken with meticulous attention to the potential environmental consequences they may pose. It is our responsibility as engineers to ensure that these projects are developed in a manner that mitigates potential adverse effects on the environment.

One of the primary environmental concerns in coastal and offshore engineering is the impact on marine life. The construction and operation of offshore structures can disrupt habitats, alter the behavior of marine organisms, and potentially lead to the loss of biodiversity. Therefore, it is necessary to conduct thorough environmental impact assessments before initiating any project. By adopting innovative technologies and design approaches, engineers can minimize the impact on marine ecosystems and ensure the protection of vulnerable species.

Another critical consideration is the potential for sediment transport and coastal erosion. Coastal and offshore engineering projects can significantly alter sediment transport patterns, leading to erosion or sediment accumulation in different areas. Understanding the coastal dynamics and sedimentation processes is essential to mitigate the potential negative consequences of these projects. By implementing appropriate sediment management strategies, engineers can maintain the stability of coastlines and protect valuable coastal ecosystems.

Furthermore, the construction and operation of coastal and offshore structures may have implications for water quality and marine pollution. The release of pollutants during construction or accidental spills can have adverse effects on marine organisms and ecosystems. Engineers must prioritize the use of environmentally friendly materials, develop robust pollution prevention plans, and establish effective monitoring systems to ensure water quality standards are upheld.

In conclusion, as engineers specializing in geological engineering, it is crucial to recognize the environmental considerations in coastal and offshore engineering. By conducting thorough environmental impact assessments, adopting innovative technologies, and implementing appropriate management strategies, we can ensure that marine renewable energy systems and other projects are developed sustainably. By striking a balance between harnessing renewable energy and preserving our natural ecosystems, we can contribute to a sustainable future for generations to come.

Chapter 3: Design and Management of Coastal Structures

Breakwaters and Seawalls

Introduction

In the field of marine renewable energy systems, engineers often encounter challenges related to the interaction between structures and the marine environment. Breakwaters and seawalls are essential engineering solutions that play a critical role in coastal protection and the sustainable development of marine renewable energy systems. This subchapter aims to provide engineers, particularly those in the niche of geological engineering, with an overview of breakwaters and seawalls, their designs, and their importance in marine renewable energy systems.

Design Considerations

The design of breakwaters and seawalls requires a thorough understanding of geological and oceanographic conditions. Geological engineers must assess the stability of the coastal area, sediment transport, and wave climate to determine the appropriate design parameters. Factors such as wave height, period, direction, and the characteristics of the seabed all influence the design process.

Breakwaters

Breakwaters are structures designed to reduce wave energy and protect coastal areas from erosion. They can be classified into two main types: rubble mound breakwaters and vertical breakwaters. Engineers must consider the specific objectives of the breakwater, the local wave climate, and the available construction materials when selecting the most appropriate design.

Seawalls

Seawalls, on the other hand, are vertical structures built parallel to the shoreline to provide protection against wave action. They are typically constructed from materials such as concrete or steel and must be designed to withstand the forces exerted by waves, tides, and storm surges. Geological engineers play a crucial role in determining the appropriate design based on the geological conditions and expected wave loads.

Integration with Marine Renewable Energy Systems

Breakwaters and seawalls are not only vital for coastal protection but also have the potential to enhance the development of marine renewable energy systems. Breakwaters, for instance, can act as wave energy converters by harnessing the energy from incoming waves. By incorporating wave energy converters into the design of breakwaters, engineers can optimize the use of coastal areas for both protection and energy generation.

Conclusion

In conclusion, breakwaters and seawalls are essential engineering solutions in the field of marine renewable energy systems. Geological engineers play a crucial role in designing these structures by considering geological conditions, wave climate, and the objectives of coastal protection. Additionally, the integration of wave energy converters into breakwater designs presents exciting opportunities for sustainable energy generation. By understanding and implementing effective designs, engineers can contribute to the development of a sustainable future in marine renewable energy systems.

Revetments and Groynes

Revetments and Groynes in Marine Renewable Energy Systems

Introduction:

In the field of marine renewable energy systems, engineers often encounter challenges related to the stability and long-term performance of structures in coastal environments. Geological engineers play a crucial role in understanding the geology of coastlines and developing effective solutions to mitigate erosion and maintain the integrity of marine renewable energy systems. This subchapter explores the concepts of revetments and groynes as engineered solutions to address these challenges.

Revetments:

Revetments are protective structures designed to stabilize shorelines against erosion caused by wave action, currents, and tides. Geological engineers must carefully consider the geological characteristics of the coastal area, such as sediment type, wave energy, and water level fluctuations, to develop appropriate revetments. Common types of revetments include concrete walls, rock armoring, and geotextile bags filled with sand or other materials. These structures act as barriers, dissipating wave energy and reducing erosion.

Geological engineers must account for various factors when designing revetments, such as the angle of slope, material selection, and potential impacts on the local ecosystem. A detailed understanding of the coastal geomorphology is essential to ensure the revetments are compatible with the natural processes occurring in the area. By implementing revetments, engineers can protect marine renewable energy systems from coastal erosion, ensuring their long-term stability.

Groynes:

Groynes are structures built perpendicular to the shoreline to interrupt the longshore transport of sediment. They trap sediment and prevent its movement along the coast, creating a wider beach and reducing erosion. Geological engineers play a vital role in determining the optimal spacing, length, and orientation of groynes based on the local coastal processes and sediment characteristics.

Properly designed groynes can significantly reduce sediment loss, minimizing the impact of erosion on marine renewable energy systems. However, engineers must consider potential drawbacks, such as changes in sedimentation patterns and impacts on neighboring coastal areas. A comprehensive understanding of coastal dynamics is crucial to ensure that groynes effectively stabilize the shoreline without causing adverse effects elsewhere.

Conclusion:

In summary, revetments and groynes are valuable tools in the arsenal of geological engineers working on marine renewable energy systems. By designing and implementing these structures with careful consideration of the geological characteristics and coastal processes, engineers can protect offshore energy installations from erosion, ensuring their sustainable performance. The expertise of geological engineers is essential in developing effective and environmentally friendly solutions to mitigate the challenges posed by coastal environments.

Beach Nourishment and Sediment Management

Introduction:

In the field of marine renewable energy systems, engineers play a critical role in designing sustainable solutions that harness the power of the ocean. One crucial aspect of this process is understanding beach nourishment and sediment management techniques. In this subchapter, we will explore the significance of beach nourishment and sediment management for engineers, with a specific focus on the niche of geological engineering.

Understanding Beach Nourishment:

Beach nourishment refers to the process of replenishing eroded beaches with sediment, usually in the form of sand. Geological engineers need to understand the complex dynamics of coastal erosion and sediment transport to effectively implement beach nourishment strategies. By studying the geological makeup of the coastline, these engineers can identify potential erosion hotspots and determine the appropriate amount and type of sediment required for nourishment.

Benefits of Beach Nourishment:

Beach nourishment offers numerous benefits for coastal areas. Firstly, it helps protect infrastructure such as buildings, roads, and utilities from the damaging effects of erosion. By adding sediment to eroded beaches, engineers create a buffer zone that absorbs wave energy, reducing the impact on the coastline. Additionally, nourished beaches provide recreational and aesthetic value, attracting tourists and boosting local economies.

Sediment Management Techniques:

Effective sediment management is crucial for maintaining the stability and ecological health of coastal areas. Geological engineers utilize various techniques to manage sediment, such as dredging, beach scraping, and sediment bypassing. Dredging involves removing sediment from one area and depositing it in another, while beach scraping involves redistributing sediment along the shoreline. Sediment bypassing refers to the redirection of sediment around man-made structures, ensuring a continuous supply to adjacent beaches.

Challenges and Considerations:

Engineers in the field of geological engineering must also address challenges associated with beach nourishment and sediment management. These include understanding the impacts of nourishment on local ecosystems, ensuring compatibility with existing coastal infrastructure, and considering the long-term sustainability of sediment sources. Additionally, engineers need to consider the potential effects of climate change on coastal erosion rates and adapt their strategies accordingly.

Conclusion:

Beach nourishment and sediment management are vital components of marine renewable energy systems, particularly for geological engineers. By understanding the dynamics of coastal erosion, engineers can design effective strategies to replenish eroded beaches and manage sediment. This subchapter has provided an overview of beach nourishment techniques, the benefits they offer, and the challenges associated with sediment management. By incorporating these considerations into their designs, engineers can contribute to a sustainable future for coastal communities and marine renewable energy systems.

Chapter 4: Design and Management of Offshore Platforms

Fixed Offshore Platforms

Fixed offshore platforms play a crucial role in the development and operation of marine renewable energy systems. These platforms provide a stable foundation for various devices, such as wind turbines, wave energy converters, and tidal turbines, that harness the power of the ocean to generate clean and sustainable energy. As engineers specializing in geological engineering, understanding the design, construction, and maintenance of fixed offshore platforms is essential to ensure the long-term viability and safety of marine renewable energy systems.

Designing fixed offshore platforms involves a multidisciplinary approach that combines principles from civil, mechanical, and geological engineering. Geological engineers play a vital role in assessing the seabed conditions and understanding the geological characteristics of the site where the platform will be installed. This information is crucial for determining the most suitable foundation design and construction techniques that can withstand the harsh marine environment.

The subchapter on fixed offshore platforms will delve into the various types of foundations commonly used, including monopiles, gravity-based structures, and suction caissons. Each foundation type has its advantages and limitations, and a thorough understanding of their behavior under different environmental conditions is necessary for successful platform design. The subchapter will also explore the geotechnical considerations for foundation design, including soil-structure interaction, bearing capacity, and settlement analysis.

Furthermore, the subchapter will discuss the construction techniques involved in building fixed offshore platforms. It will cover topics such as installation methods, offshore construction equipment, and the challenges faced during the construction phase. Special emphasis will be placed on the importance of geotechnical surveys and site investigations to ensure the platform's integrity and stability over its operational lifespan.

Maintenance and inspection of fixed offshore platforms are vital to ensure the continuous and reliable operation of marine renewable energy systems. The subchapter will outline the various inspection techniques used to assess the structural integrity of the platform, including underwater inspections, remote sensing technologies, and structural health monitoring systems. It will also discuss the maintenance strategies employed to mitigate corrosion, fatigue, and other degradation mechanisms that can affect the platform's performance.

In conclusion, the subchapter on fixed offshore platforms is aimed at providing engineers specializing in geological engineering with a comprehensive understanding of the design, construction, and maintenance aspects associated with these critical structures. By gaining expertise in this field, engineers can contribute to the development of sustainable marine renewable energy systems and help shape a greener future for our planet.

Floating Offshore Platforms

In the realm of marine renewable energy systems, floating offshore platforms play a crucial role in harnessing the power of the ocean to generate sustainable energy. Designed to withstand the harsh and dynamic marine environment, these platforms offer a promising solution for the growing demand for clean and renewable energy sources. This subchapter aims to provide engineers specializing in geological engineering with a comprehensive understanding of floating offshore platforms and their significance in marine renewable energy systems.

Floating offshore platforms are innovative structures that allow for the installation of various types of renewable energy devices, such as wind turbines, tidal turbines, and wave energy converters, in deep waters where fixed structures are not feasible. These platforms are typically anchored to the seabed using mooring systems, which ensure stability and provide support for the energy devices. The choice of mooring system depends on factors such as water depth, wave and wind conditions, and the specific requirements of the renewable energy device.

Geological engineering plays a crucial role in the design and installation of floating offshore platforms. Understanding the seabed conditions, including soil type, composition, and stability, is essential for optimizing the mooring system and ensuring the long-term integrity of the platform. Geotechnical investigations, including site surveys and soil testing, are conducted to assess the soil properties and determine the most suitable anchoring technique.

The subchapter will delve into the various types of floating offshore platforms, including tension leg platforms (TLPs), spar platforms, semi-submersible platforms, and floating wind turbines. Each type has its own advantages and challenges, and engineers need to consider factors such as water depth, wave climate, wind conditions, and environmental impacts when selecting the appropriate platform for a specific location.

Furthermore, the subchapter will discuss the structural design and materials used in floating offshore platforms to ensure robustness and durability in harsh marine environments. It will cover topics such as fatigue analysis, dynamic response, and corrosion protection, providing engineers with the necessary knowledge to design platforms that can withstand extreme weather conditions and sea states.

Overall, this subchapter aims to equip engineers specializing in geological engineering with the knowledge and understanding required to design, analyze, and install floating offshore platforms for marine renewable energy systems. By providing insights into the various types of platforms, geotechnical considerations, and structural design aspects, this subchapter serves as a valuable resource for engineers involved in the sustainable future of marine renewable energy.

Subsea Structures and Foundations

In the quest for sustainable energy sources, marine renewable energy systems have emerged as a promising solution. These systems harness the power of the ocean to generate electricity, and their successful implementation relies heavily on the design and construction of robust subsea structures and foundations. This subchapter explores the critical aspects of subsea structures and foundations, highlighting the role of geological engineering in ensuring their integrity and resilience.

Subsea structures serve as the backbone of marine renewable energy systems, providing support and stability to various components such as turbines, wave energy converters, and tidal energy devices. These structures must withstand extreme hydrodynamic forces, corrosive environments, and unpredictable seabed conditions. Geological engineers play a vital role in assessing the geological characteristics of the seabed and designing suitable foundations that can withstand these challenges.

One of the key considerations in subsea structure design is the selection of appropriate foundation types. Geological engineers analyze the seabed composition, including soil properties, rock formations, and sedimentary layers, to determine the most suitable foundation option. Common foundation types include gravity-based structures, piles, suction caissons, and suction buckets. Each type has its advantages and limitations, and geological engineers must carefully evaluate the site-specific conditions to make informed decisions.

The subchapter also explores the importance of geotechnical investigations in subsea structure design. Geological engineers conduct detailed site investigations to gather data on the seabed conditions, including soil strength, stability, and liquefaction potential.

This information is crucial for designing foundations that can withstand the dynamic loads imposed by ocean waves, currents, and tides. Various geotechnical testing methods, such as cone penetration tests, pressuremeter tests, and seismic surveys, are employed to assess the soil properties accurately.

Furthermore, the subchapter delves into the challenges associated with subsea structure installation and maintenance. Geological engineers must consider the complex interaction between the structure, the foundation, and the seabed during installation to ensure stability and prevent potential damage. Regular inspections and maintenance are also essential to monitor the integrity of subsea structures and address any issues promptly.

Overall, this subchapter emphasizes the significance of geological engineering in the design, construction, and maintenance of subsea structures and foundations for marine renewable energy systems. By understanding the geological characteristics of the seabed and employing suitable foundation types, engineers can ensure the longevity and reliability of these structures, contributing to the sustainable future of marine renewable energy.

Chapter 5: Introduction to Marine Renewable Energy Systems

Overview of Renewable Energy Sources

Renewable energy sources are gaining increasing attention and significance in today's world as the need for sustainable energy solutions becomes more pressing. This subchapter will provide an overview of the various renewable energy sources available, with a specific focus on their applications within the field of geological engineering.

Geological engineering encompasses the study and application of geological principles to engineering projects, including the exploration and extraction of natural resources, environmental protection, and infrastructure development. With a growing emphasis on sustainability, geotechnical engineers are increasingly seeking renewable energy sources that align with their field's core principles.

One prominent renewable energy source relevant to geological engineering is solar energy. Solar power harnesses the energy from the sun and converts it into electricity through the use of photovoltaic cells. This energy source is particularly well-suited for geological engineers working in remote locations or areas with abundant sunlight. Solar energy can be used to power geotechnical equipment, such as drilling rigs, monitoring devices, and data collection systems, thereby reducing reliance on fossil fuels and minimizing environmental impact.

Another renewable energy source of interest to geological engineers is wind energy. Wind power utilizes the kinetic energy from the wind to generate electricity through wind turbines. Geological engineers can incorporate wind energy systems into their projects, especially in coastal or mountainous regions where wind resources are abundant. Wind turbines can be used to power various geotechnical activities, such as soil and rock sampling, geophysical surveys, and water pumping for construction purposes.

Hydropower, derived from the gravitational force of falling or flowing water, is yet another renewable energy source relevant to geological engineering. Geotechnical engineers often work in close proximity to water bodies, such as rivers, lakes, and oceans, making hydropower an attractive option. By integrating hydropower systems into their projects, engineers can harness the energy generated from water flow to power construction equipment, water management systems, and other geotechnical operations.

Geothermal energy, which utilizes the heat from within the earth's crust, is also worth considering for geological engineers. This renewable energy source can provide heating and cooling solutions for various geotechnical applications, including ground-source heat pumps, geothermal drilling operations, and underground storage facilities.

In conclusion, this subchapter has provided an overview of renewable energy sources within the context of geological engineering. Solar energy, wind power, hydropower, and geothermal energy offer sustainable alternatives to traditional fossil fuel-based energy sources. By incorporating these renewable energy systems into their projects, geological engineers can contribute to a more sustainable future while ensuring the continued development and progress of their field.

Types of Marine Renewable Energy Systems

Marine renewable energy systems are becoming increasingly popular as the world seeks sustainable solutions to meet its growing energy demands. These systems harness the power of the ocean's waves, tides, and currents to generate electricity, offering a promising alternative to fossil fuels. Among the various types of marine renewable energy systems, the following have gained significant attention in recent years.

1. Wave Energy Converters (WECs): Wave energy converters capture the kinetic energy from ocean waves and convert it into electricity. There are several types of WECs, including point absorbers, attenuators, and oscillating water columns. Point absorbers move up and down with the waves, converting the mechanical energy into electricity. Attenuators, on the other hand, are long structures that move with the waves and generate power through the relative motion between their segments. Oscillating water columns use the rise and fall of water levels inside a chamber to drive an air turbine, producing electricity.

2. Tidal Energy Systems: Tidal energy systems employ turbines or underwater devices to capture the energy from tidal currents and convert it into electricity. There are two main types of tidal energy systems: tidal stream and tidal barrage. Tidal stream systems use underwater turbines that are driven by the kinetic energy of the tides. These turbines can be mounted on the seabed or floating structures. Tidal barrage systems, on the other hand, rely on the potential energy created by the difference in water levels between high and low tides. This energy is harnessed by constructing barrages that contain turbines, which generate electricity as the water flows in and out.

3. Ocean Thermal Energy Conversion (OTEC): OTEC systems utilize the temperature difference between warm surface water and cold deep water to generate electricity. This temperature gradient is exploited through the use of heat exchangers and a working fluid (such as ammonia) that vaporizes at low temperatures. The vaporized fluid drives a turbine, producing electricity. OTEC systems can be either closed-cycle or open-cycle, depending on the working fluid used.

These marine renewable energy systems offer immense potential for a sustainable future. However, their deployment and integration into existing energy grids require significant engineering solutions. Geological engineers play a crucial role in assessing the seabed conditions, identifying suitable locations for installations, and ensuring the structural integrity of marine renewable energy systems. By understanding the various types of marine renewable energy systems, engineers can contribute to the development of sustainable energy solutions that minimize environmental impact while meeting the world's growing energy needs.

Advantages and Disadvantages of Marine Renewable Energy

Marine renewable energy, also known as ocean energy, is a promising field that harnesses the power of the sea to generate electricity. It offers a sustainable alternative to fossil fuels and has the potential to significantly reduce greenhouse gas emissions. However, like any other energy source, marine renewable energy has its own set of advantages and disadvantages. In this subchapter, we will explore these pros and cons, specifically from the perspective of engineers specializing in geological engineering.

Advantages:

1. Abundant and Renewable: The ocean is an inexhaustible resource, with vast amounts of potential energy waiting to be tapped. Tides, waves, and currents are constantly generated by natural processes, ensuring a continuous supply of renewable energy.

2. Predictable and Reliable: Unlike other renewable energy sources such as solar and wind, marine energy is highly predictable. Tidal cycles and wave patterns can be accurately forecasted, allowing engineers to plan and optimize energy production. This reliability makes it easier to integrate marine energy into the existing power grid.

3. Low Carbon Footprint: Marine renewable energy systems produce electricity without emitting greenhouse gases or other harmful pollutants. This makes them a clean and eco-friendly alternative to conventional fossil fuel-based power generation, contributing to the reduction of global carbon emissions.

Disadvantages:

1. High Initial Costs: Developing and installing marine energy infrastructure can be expensive. The construction of offshore wind

farms, tidal turbines, and wave energy converters requires significant investment in research, development, and deployment. These costs can be a barrier to widespread adoption.

2. Environmental Impact: While marine renewable energy is generally considered environmentally friendly, there are potential negative impacts to consider. For example, the installation and operation of tidal turbines or wave energy devices can disrupt marine ecosystems, affecting marine life and habitats. Proper environmental assessments and monitoring are crucial to mitigate these impacts.

3. Limited Technology Maturity: Compared to other renewable energy sources such as solar and wind, marine energy technologies are still in the early stages of development. This means that there may be uncertainties and challenges associated with the reliability, efficiency, and maintenance of marine energy systems. Continued research and innovation are needed to overcome these barriers.

In conclusion, marine renewable energy presents numerous advantages for engineers specializing in geological engineering. Its abundance, predictability, and low carbon footprint make it an attractive option for sustainable energy generation. However, the high initial costs, potential environmental impacts, and technological challenges are factors that need to be carefully assessed and managed. With ongoing advancements and increased investment, marine renewable energy has the potential to play a significant role in our transition towards a more sustainable future.

Chapter 6: Wave Energy Conversion Systems

Wave Characteristics and Measurement

Introduction

In the field of marine renewable energy, understanding wave characteristics and measurement is crucial for engineers specializing in geological engineering. Waves are a powerful natural resource that can be harnessed to generate sustainable energy. To effectively design and implement marine renewable energy systems, engineers must have a thorough understanding of the various characteristics of waves and the methods used to measure them. This subchapter aims to provide an overview of wave characteristics and measurement techniques for engineers in the field of geological engineering.

Wave Characteristics

Waves possess several key characteristics that engineers must consider when designing marine renewable energy systems. These characteristics include wave height, period, wavelength, and amplitude. Wave height refers to the vertical distance between the crest and trough of a wave, while period represents the time taken for one complete wave cycle. Wavelength, on the other hand, is the horizontal distance between two consecutive wave crests or troughs. Finally, wave amplitude refers to the maximum displacement of particles from their rest position.

Measurement Techniques

Accurate measurement of wave characteristics is essential for the successful implementation of marine renewable energy systems. Engineers rely on various measurement techniques to obtain reliable data. One commonly used method is wave buoys, which are equipped with sensors to measure wave height, period, and direction. These buoys are deployed at strategic locations to collect real-time data about wave behavior.

Another technique employed by engineers is the use of pressure sensors on the seabed. These sensors measure the changes in water pressure caused by passing waves, allowing engineers to calculate wave characteristics. Acoustic Doppler Current Profilers (ADCPs) are also commonly used to measure waves by analyzing the Doppler shift in the frequency of sound waves reflected off particles in the water column.

Conclusion

For engineers specializing in geological engineering, understanding wave characteristics and measurement techniques is crucial for the successful implementation of marine renewable energy systems. Waves possess unique characteristics such as height, period, wavelength, and amplitude, which must be accurately measured. Engineers employ various techniques such as wave buoys, pressure sensors, and ADCPs to collect reliable data on wave behavior. This knowledge aids engineers in designing efficient and sustainable marine renewable energy systems that harness the power of waves to generate electricity. By prioritizing the understanding and measurement of wave characteristics, engineers can contribute to the development of a sustainable future in the field of marine renewable energy.

Wave Energy Conversion Technologies

Wave energy conversion technologies are innovative engineering solutions that harness the power of ocean waves to generate renewable energy. As the demand for sustainable energy sources continues to grow, engineers in the field of geological engineering are actively researching and developing methods to harness the abundant energy potential of the world's oceans.

This subchapter provides an overview of various wave energy conversion technologies, highlighting their engineering principles, advantages, and challenges. By understanding these technologies, engineers specializing in geological engineering can contribute to the design and implementation of efficient and reliable wave energy conversion systems.

One of the most promising wave energy conversion technologies is the Oscillating Water Column (OWC). The OWC is a device that captures wave energy by using the rise and fall of water levels in a chamber to drive an air turbine. This turbine then generates electricity. The OWC has several advantages, including its simplicity, low maintenance requirements, and the ability to operate in a wide range of wave conditions. However, challenges such as the need for large-scale structures and the impact of extreme weather conditions on the system's performance need to be addressed for widespread deployment.

Another technology gaining attention in the field of wave energy conversion is the Point Absorber. This device consists of a buoyant structure that moves up and down with the waves, driving a generator to produce electricity. Point absorbers are particularly suitable for locations where wave conditions vary significantly. Engineers are working on optimizing the design of these devices to maximize energy extraction and improve their efficiency.

In addition to these technologies, engineers are also exploring other innovative concepts, such as the Wave Dragon, which uses a large floating platform to capture and convert wave energy, and the Attenuator, which consists of interconnected segments that convert the wave's energy into electricity. These technologies show great promise but require further research and development to become commercially viable.

As geological engineers, understanding the characteristics of the ocean waves and the geology of coastal areas is crucial for the successful implementation of wave energy conversion technologies. By utilizing their expertise in geological engineering, engineers can contribute to site selection, foundation design, and environmental impact assessments, ensuring the sustainable and responsible deployment of wave energy conversion systems.

In conclusion, wave energy conversion technologies offer immense potential for generating renewable energy from the ocean's waves. Engineers specializing in geological engineering play a vital role in developing and implementing these technologies, considering the unique geological aspects of coastal areas. Through continuous research and innovation, engineers can contribute to the realization of a sustainable future powered by wave energy.

Design and Optimization of Wave Energy Converters

Wave energy converters (WECs) are devices that harness the power of ocean waves and convert it into usable energy. These devices play a critical role in marine renewable energy systems, as they provide a sustainable and clean source of power. In this subchapter, we will explore the design and optimization of WECs, focusing on the specific considerations and challenges that engineers, particularly those specializing in geological engineering, need to address.

The design of WECs involves understanding the dynamics of ocean waves and developing efficient systems to capture their energy. Geological engineers play a crucial role in this process, as they possess the expertise to assess the seabed conditions, wave characteristics, and coastal geology, which directly impact the design and placement of WECs. By analyzing geological data and conducting site-specific assessments, engineers can determine the optimal locations for installing WECs, maximizing their energy capture potential.

Furthermore, optimization techniques are essential to enhance the performance and efficiency of WECs. Engineers must consider various factors, such as wave height, period, direction, and the device's response to these parameters. By employing advanced modeling and simulation tools, geological engineers can evaluate different design configurations and optimize the energy conversion process. This involves analyzing the structural integrity, power generation capacity, and overall system performance of WECs under varying wave conditions.

In addition to design and optimization, engineers need to address the challenges associated with deploying and maintaining WECs. Geological engineers are uniquely equipped to assess the potential environmental impacts of WEC installations, including seabed disturbances, changes in coastal sediment transport, and potential effects on marine ecosystems. By considering these factors during the design phase, engineers can mitigate any adverse effects and ensure sustainable energy production.

Overall, the design and optimization of WECs require a multidisciplinary approach, with geological engineering playing a pivotal role. By integrating geological expertise with engineering principles, engineers can develop efficient and environmentally sustainable solutions for harnessing wave energy. This subchapter will delve into the technical aspects of WEC design, optimization techniques, and the geological considerations that engineers specializing in geological engineering need to address. It will provide valuable insights and guidance for professionals and researchers working in marine renewable energy systems, specifically in the niche of geological engineering.

Chapter 7: Tidal Energy Conversion Systems

Tidal Characteristics and Measurement

Introduction:

Tides, the rise and fall of sea levels caused by the gravitational forces of the Moon and the Sun, play a crucial role in the field of marine renewable energy systems. Understanding tidal characteristics and accurately measuring them are essential for engineers, particularly those specializing in geological engineering, to design and optimize sustainable solutions for harnessing tidal energy. This subchapter delves into the various aspects of tidal characteristics and the methods employed in their measurement.

Tidal Characteristics:

The study of tidal characteristics involves analyzing the amplitude, period, and phase of tides. Engineers specializing in geological engineering must comprehend the factors influencing these characteristics, such as lunar and solar gravitational forces, topography, and oceanic basin shape. Furthermore, tidal patterns and variations, including spring and neap tides, need to be considered during the design and implementation of marine renewable energy systems.

Measurement Techniques:

Accurate measurement of tidal characteristics is crucial for the successful development of marine renewable energy projects. Engineers working in geological engineering employ various techniques to measure tidal parameters. These techniques include tidal gauges, acoustic doppler current profilers (ADCPs), and remote sensing technologies such as satellite altimetry. Each method has its advantages and limitations, and a combination of these techniques is often employed to obtain comprehensive tidal data.

Tidal Gauge:

Tidal gauges, also known as tide gauges or sea-level recorders, are widely used to measure tidal heights and variations. These devices consist of a sensor connected to a recording system that continuously records the water level. Engineers can use tidal gauge data to determine tidal characteristics and create tidal prediction models.

Acoustic Doppler Current Profilers (ADCPs):

ADCPs are commonly used to measure water velocities and currents in tidal environments. These instruments utilize sound waves to measure the Doppler shift caused by moving particles in the water. By deploying ADCPs at various locations, engineers can gather data on tidal currents and their variations, aiding in the design and optimization of marine renewable energy systems.

Remote Sensing Technologies:

Remote sensing technologies, such as satellite altimetry, provide a broader perspective on tidal characteristics. These techniques enable engineers to collect data on global tidal patterns and variations over large spatial scales. By combining satellite altimetry data with other measurement techniques, engineers can gain a comprehensive understanding of tidal characteristics, supporting the development of sustainable marine renewable energy solutions.

Conclusion:

Tidal characteristics and their accurate measurement are vital for engineers specializing in geological engineering to design and optimize marine renewable energy systems. By understanding tidal patterns, variations, and the factors influencing them, engineers can develop sustainable solutions for harnessing tidal energy. Tidal gauges, ADCPs, and remote sensing technologies are among the measurement techniques employed to gather comprehensive tidal data. The combination of these methods enables engineers to obtain crucial information for the successful implementation of marine renewable energy projects.

Tidal Energy Conversion Technologies

The harnessing of tidal energy has gained significant attention in recent years due to its potential as a reliable and sustainable source of renewable energy. This subchapter will explore the various technologies used for the conversion of tidal energy into electricity, providing engineers in the field of geological engineering with a comprehensive understanding of the available options.

1. Tidal Barrages: Tidal barrages are large dams constructed across estuaries or bays, utilizing the potential energy of the tidal range to generate electricity. As the tide rises, water is allowed to flow into the basin through turbines, which generate electricity. During ebb tide, the water is released back into the sea. Tidal barrages offer high energy conversion efficiency but can have significant environmental impacts due to changes in water flow patterns and habitat disturbances.

2. Tidal Stream Turbines: Tidal stream turbines operate similarly to wind turbines, but instead of wind, they harness the kinetic energy of tidal currents. These turbines are typically installed on the seabed or mounted on floating platforms, and as the tidal currents flow, the rotating blades of the turbine generate electricity. Tidal stream turbines are relatively easy to install and maintain, and their impact on the environment is generally minimal.

3. Dynamic Tidal Power (DTP): DTP is a relatively new concept that involves constructing a long dam-like structure perpendicular to the coastline, forming a semi-enclosed basin. As the tidal currents flow through the structure, the water level within the basin rises, creating a potential energy gradient. By strategically opening and closing gates, the water is released, driving turbines to generate electricity. DTP has

the advantage of being able to generate power from both ebb and flood tides, maximizing energy output.

4. Tidal Kites: Tidal kites, also known as underwater kites, are a promising emerging technology in tidal energy conversion. These kite-like structures are tethered to the seabed and use the movement of tidal currents to generate electricity. As the kite moves in a figure-eight pattern, the tether is spooled out and retracted, driving a generator. Tidal kites offer high power generation potential and have minimal environmental impact.

In conclusion, tidal energy conversion technologies offer a range of options for engineers in the field of geological engineering to harness the power of the tides for sustainable energy production. Each technology has its advantages and considerations, and the choice of technology will depend on site-specific conditions, environmental impacts, and economic feasibility. With ongoing advancements in technology and research, tidal energy is poised to play a significant role in the transition towards a sustainable future.

Design and Optimization of Tidal Energy Converters

The design and optimization of tidal energy converters play a crucial role in harnessing the immense power of the oceans and providing sustainable energy solutions for the future. As engineers in the field of geological engineering, understanding the principles behind these devices and their efficient utilization is essential.

Tidal energy converters are devices that capture the kinetic energy from the ebb and flow of tides and convert it into electricity. The design process involves various considerations, including the selection of appropriate sites, understanding the tidal patterns, and optimizing the device's performance.

One of the key factors in designing tidal energy converters is the selection of the site. Geological engineers play a vital role in evaluating the geological characteristics of potential locations. Factors such as water depth, seabed composition, and tidal range impact the choice of device and its installation method. By analyzing geological data, engineers can determine the best locations for maximum energy extraction and minimal environmental impact.

Once the site is chosen, engineers focus on optimizing the device's performance. This involves designing efficient hydrodynamic shapes that capture the tidal energy effectively. Numerical modeling and simulation techniques are employed to understand the complex fluid dynamics involved. By studying the flow patterns and forces acting on the device, engineers can refine the design to maximize power output and minimize structural loads.

Another aspect of optimization is the choice of materials and construction techniques. Geological engineers play a critical role in selecting materials that can withstand the harsh marine environment, including corrosion and fatigue. They also evaluate the feasibility of installation methods, considering factors such as seabed stability and the impact on marine ecosystems.

In addition to the design and optimization process, engineers must also consider the environmental impact of tidal energy converters. Geological engineers, with their expertise in environmental systems, can assess the potential effects on marine life, sediment transport, and coastal erosion. By integrating environmental considerations into the design process, engineers can ensure sustainable and responsible development of tidal energy projects.

In conclusion, the design and optimization of tidal energy converters require a multidisciplinary approach, with geological engineering playing a crucial role. By understanding the geological characteristics of potential sites, optimizing device performance, and considering environmental impacts, engineers can contribute to the development of sustainable and efficient tidal energy systems. With their expertise, engineers in the field of geological engineering are poised to shape the future of marine renewable energy systems.

Chapter 8: Ocean Current Energy Conversion Systems

Ocean Current Characteristics and Measurement

Introduction:

In the field of marine renewable energy systems, understanding ocean current characteristics and accurately measuring them is essential for successful engineering solutions. This subchapter will delve into the key characteristics of ocean currents and explore various methods for their measurement. Specifically tailored for engineers in the niche of geological engineering, this content aims to provide a comprehensive overview of the subject.

Ocean Current Characteristics:

Ocean currents are continuous movements of water in the ocean, driven by various factors such as temperature, salinity, wind, and the Earth's rotation. These currents can be classified into two main types: surface currents and deep ocean currents. Surface currents are driven primarily by wind and can extend up to a few hundred meters, while deep ocean currents are influenced by density differences and can reach depths of several kilometers.

Measurement Techniques:

Accurate measurement of ocean currents is crucial for the design and implementation of marine renewable energy systems. Engineers in geological engineering must understand the various techniques available for this purpose. One commonly used method is acoustic Doppler current profilers (ADCPs), which utilize sound waves to measure water velocity at different depths. Another technique is the use of current meters, which can be deployed on buoys or moorings to measure current speed and direction.

Other important measurement techniques include satellite remote sensing, where satellites equipped with radar or altimeters provide information on surface currents, and numerical models that simulate ocean dynamics based on inputs such as temperature, salinity, and wind data. These models help predict and understand the behavior of ocean currents, aiding engineers in developing sustainable solutions.

Challenges and Future Developments: While significant progress has been made in measuring ocean currents, there are still challenges to overcome. These include the high costs associated with deploying instruments in remote and harsh marine environments, the need for long-term data collection, and the complexity of accurately capturing small-scale features.

However, advancements in technology offer promising solutions. Miniaturized sensors and autonomous platforms like underwater gliders have the potential to revolutionize ocean current measurements, providing cost-effective and real-time data collection. Additionally, advancements in data assimilation techniques and high-resolution models will further enhance our understanding of ocean currents.

Conclusion:
Ocean current characteristics and measurement are vital components in the field of marine renewable energy systems. Engineers specializing in geological engineering need to stay abreast of the latest techniques and technologies for accurately measuring ocean currents. By understanding the characteristics and behavior of these currents,

engineers can design sustainable and efficient marine renewable energy systems that harness the vast potential of our oceans.

Ocean Current Energy Conversion Technologies

In recent years, the exploration and development of marine renewable energy systems have gained significant attention due to their potential as sustainable alternatives to traditional fossil fuel-based energy sources. Among these systems, ocean current energy conversion technologies have emerged as one of the most promising solutions for harnessing the immense power of ocean currents to generate electricity. This subchapter will delve into the various ocean current energy conversion technologies, providing engineers in the niche of geological engineering with valuable insights and engineering solutions for a sustainable future.

Ocean currents, driven by a combination of wind, temperature, salinity, and Earth's rotation, possess vast amounts of kinetic energy that can be converted into usable electricity. The subchapter will begin by discussing the two main types of ocean current energy conversion technologies: tidal stream turbines and ocean current turbines.

Tidal stream turbines, also known as underwater windmills, resemble wind turbines but are specifically designed to operate in fast-flowing tidal currents. Engineers will gain an understanding of the design considerations, including rotor size, blade pitch control, and anchoring systems, necessary to optimize the efficiency and reliability of tidal stream turbines. Additionally, the subchapter will explore the environmental impact assessment methodologies and mitigation measures associated with deploying these technologies in marine environments.

Ocean current turbines, on the other hand, are designed to extract energy from continuous ocean currents. Engineers will be introduced to the different types of ocean current turbines, such as horizontal axis turbines and vertical axis turbines, and their respective advantages and disadvantages. The subchapter will delve into the engineering challenges faced in designing robust marine structures that can withstand the harsh oceanic conditions while effectively capturing energy from ocean currents. It will also highlight the importance of accurate resource assessment and site selection to optimize energy production.

Furthermore, the subchapter will discuss the electrical and grid integration aspects of ocean current energy conversion technologies. Engineers will gain insights into power conversion systems, transmission technologies, and grid integration strategies to ensure smooth integration into existing electrical infrastructure.

By providing a comprehensive overview of ocean current energy conversion technologies, this subchapter will equip engineers specializing in geological engineering with the necessary knowledge and tools to contribute to the development and implementation of sustainable marine renewable energy systems. With a focus on engineering solutions, it will inspire and empower engineers to drive the transition towards a more sustainable future.

Design and Optimization of Ocean Current Energy Converters

Introduction:

In recent years, the exploration of renewable energy sources has gained significant attention due to the growing concerns about climate change and the need for sustainable energy solutions. Among various renewable energy options, ocean currents have emerged as a promising resource for generating clean and reliable electricity. This subchapter explores the design and optimization of ocean current energy converters, with a focus on the engineering solutions and challenges faced in this field.

Design challenges:

The design of ocean current energy converters presents unique challenges compared to other renewable energy systems. The harsh marine environment, unpredictable currents, and the need for robust and efficient structures require careful considerations in the design process. Engineers specialized in geological engineering play a crucial role in understanding the geology and seabed conditions to ensure the stability and longevity of these structures.

Optimization strategies:

To maximize the efficiency and economic viability of ocean current energy converters, optimization strategies are essential. Engineers must consider factors such as turbine design, materials, deployment methodologies, and maintenance requirements. By optimizing these parameters, engineers can enhance the overall performance and reliability of the converters, ultimately increasing their power output and reducing operational costs.

Materials selection:

The selection of suitable materials is crucial for the longevity and performance of ocean current energy converters. Engineers in geological engineering must consider the corrosive nature of the marine environment and the presence of sediments, rocks, and other geological features. Advanced composite materials and coatings that offer high resistance to corrosion and erosion are often preferred. Additionally, engineers must assess the potential impact of marine life on the converter's materials and incorporate appropriate measures to mitigate any adverse effects.

Structural stability:

Ensuring the structural stability of ocean current energy converters is of utmost importance. Engineers must consider the dynamic forces exerted by ocean currents, waves, and extreme weather events. Detailed site assessments and geotechnical studies are crucial in determining the seabed conditions and designing appropriate foundations. Advanced numerical modeling techniques, such as finite element analysis, can help engineers optimize the structural design, ensuring safe and reliable operation.

Conclusion:

Designing and optimizing ocean current energy converters requires specialized knowledge and skills in geological engineering. With careful consideration of the unique challenges posed by the marine environment, engineers can develop efficient and reliable systems that harness the power of ocean currents. By continuously improving design and optimization strategies, the field of ocean current energy conversion holds immense potential for contributing to a sustainable future.

Chapter 9: Offshore Wind Energy Systems

Wind Characteristics and Measurement

Introduction

In the field of marine renewable energy systems, understanding the wind characteristics and accurately measuring them is paramount. Wind energy has emerged as a sustainable and viable source of power generation, making it imperative for engineers, especially those specializing in geological engineering, to have a comprehensive understanding of wind characteristics and measurement techniques. This subchapter will delve into the key aspects of wind behavior, the factors affecting wind patterns, and the methods used to measure wind speed and direction.

Wind Behavior and Factors Affecting Wind Patterns To comprehend wind characteristics, engineers must first grasp the fundamental principles governing wind behavior. Winds are generated due to the uneven heating of the Earth's surface, resulting in the movement of air from high-pressure areas to low-pressure areas. The complex interplay of various factors, including solar radiation, atmospheric pressure, temperature gradients, and geographical features, influences wind patterns. Engineers specializing in geological engineering need to consider how landforms, coastal topography, and offshore structures can modify wind flow.

Measurement Techniques for Wind Speed and Direction Accurate measurement of wind speed and direction is vital for the successful design and operation of marine renewable energy systems. Engineers rely on a range of measurement techniques and instruments to collect wind data. Anemometers, such as cup, propeller, and sonic anemometers, are commonly used to measure wind speed. These instruments employ different principles, such as rotating cups or propellers, to determine the wind speed. Wind vanes, on the other hand, provide information about wind direction by orienting themselves with the prevailing wind. Doppler radar and remote sensing technologies have also proven effective in measuring wind characteristics over large areas.

Data   Analysis   and   Visualization
Once wind data is collected, engineers can analyze and visualize the information to gain valuable insights. Statistical analysis techniques, such as wind rose plots, frequency analysis, and power spectral density analysis, help identify wind patterns, seasonal variations, and extreme wind events. Visualization tools, including Geographic Information Systems (GIS) and computational models, enable engineers to develop accurate wind resource maps, simulate wind behavior, and optimize the placement of marine renewable energy systems.

Conclusion
Understanding wind characteristics and accurately measuring wind speed and direction are essential for engineers specializing in geological engineering to design and operate marine renewable energy systems effectively. By comprehending wind behavior and the factors influencing wind patterns, engineers can develop optimized solutions for harnessing wind energy. Utilizing advanced measurement techniques and data analysis tools, engineers can gather valuable wind data and visualize it to make informed decisions. This subchapter provides a comprehensive overview of wind characteristics and measurement methods, empowering engineers in the field of geological engineering to contribute to the development of sustainable marine renewable energy systems.

Offshore Wind Turbine Technologies

In recent years, offshore wind energy has emerged as a promising solution to address the global energy demand while reducing greenhouse gas emissions. As engineers in the field of geological engineering, it is crucial to understand the various technologies associated with offshore wind turbines and their impact on marine renewable energy systems. This subchapter aims to provide you with an overview of offshore wind turbine technologies, their design considerations, and their potential benefits for sustainable energy production.

Offshore wind turbines are specially designed structures that harness wind energy to generate electricity in marine environments. Compared to onshore wind farms, offshore installations offer several advantages, including higher wind speeds, larger turbine capacities, and reduced visual impact. However, geological engineers must consider additional factors such as seabed conditions, water depth, and wave-induced loads when designing offshore wind turbine foundations.

One of the key technologies used in offshore wind turbines is the monopile foundation. This type of foundation consists of a single steel pile driven into the seabed, providing stability and support for the turbine. Monopiles are suitable for water depths up to approximately 30 meters and are commonly used in shallow coastal areas. For greater water depths, alternative technologies such as jacket foundations, gravity-based structures, or floating platforms are employed.

Another important aspect of offshore wind turbine technologies is the design and construction of turbine blades. Engineers strive to develop longer and more efficient blades to capture a higher percentage of wind energy. Advanced materials, such as carbon fiber reinforced polymers, are being utilized to reduce the weight and increase the strength of turbine blades. Additionally, innovative blade designs, such as variable pitch and aerodynamically optimized shapes, are being explored to enhance energy conversion efficiency.

Maintenance and operation of offshore wind turbines are also critical considerations. Engineers must develop strategies to access and service turbines located far from the shore. This involves utilizing specialized vessels, robotic technologies, and remote monitoring systems to ensure safe and efficient turbine operation throughout their lifetime.

Offshore wind turbines have the potential to significantly contribute to the sustainable future of energy production. As geological engineers, understanding the various offshore wind turbine technologies and their implications for marine renewable energy systems is vital. By furthering research and development in this field, engineers can contribute to the efficient utilization of offshore wind resources and the reduction of our dependence on fossil fuels.

Design and Optimization of Offshore Wind Farms

Offshore wind farms have emerged as a vital source of renewable energy, with the potential to significantly contribute to a sustainable future. This subchapter explores the design and optimization of offshore wind farms, specifically tailored for engineers in the niche of geological engineering.

To effectively design and optimize offshore wind farms, engineers need to consider various factors, including site selection, foundation design, turbine layout, and transmission infrastructure. Geological engineers play a crucial role in assessing the site's geological conditions and determining the optimal foundation design for offshore wind turbines.

Site selection is a critical step in the design process, as it directly affects the performance and profitability of the wind farm. Geological engineers employ their expertise to evaluate the seabed conditions, seabed stability, and geological hazards such as fault lines or landslides. By conducting detailed geotechnical investigations and analyzing geological data, engineers can identify suitable locations for offshore wind farms, ensuring safe and stable installations.

Foundation design is another crucial aspect of offshore wind farm optimization. Geological engineers utilize their knowledge of soil mechanics and geotechnical engineering to design foundations that can withstand harsh marine conditions. They consider factors such as soil type, bearing capacity, and structural integrity to determine the most suitable foundation type, whether it be monopiles, jackets, or floating platforms. Through advanced numerical modeling and analysis techniques, engineers can optimize the foundation design for maximum stability and reliability.

Turbine layout is an essential consideration to maximize energy production and minimize wake effects. Geological engineers collaborate with wind farm planners and layout designers to assess the impacts of geological features, such as underwater canyons or sandbanks, on the wind flow patterns. By incorporating this geological information into the layout design, engineers can optimize the turbine arrangement, minimizing wake effects and maximizing energy yield.

Transmission infrastructure is another critical component of offshore wind farm design. Geological engineers work alongside electrical engineers to assess the seabed conditions for cable routing and burial. They evaluate the geological hazards, such as scouring or sediment transport, that may affect cable integrity. By optimizing the transmission infrastructure design, engineers can ensure efficient power transmission from the wind farm to the onshore grid.

In conclusion, the design and optimization of offshore wind farms require the expertise of geological engineers. By considering factors such as site selection, foundation design, turbine layout, and transmission infrastructure, engineers can maximize the performance and profitability of offshore wind farms. Through their specialized knowledge in geological engineering, these professionals play a vital role in the sustainable development of marine renewable energy systems.

Chapter 10: Environmental Impact and Mitigation Strategies

Environmental Impact of Marine Renewable Energy Systems

As engineers in the field of geological engineering, it is crucial to have a comprehensive understanding of the environmental impact of marine renewable energy systems. In this subchapter, we will explore the various aspects of this impact and discuss the engineering solutions that can contribute to a sustainable future.

Marine renewable energy systems, such as tidal and wave energy converters, hold immense potential for generating clean and sustainable power. However, it is essential to carefully evaluate their environmental impact to ensure that these systems do not harm marine ecosystems or compromise the long-term health of our oceans.

One of the primary concerns associated with marine renewable energy systems is the potential disruption to marine life. Construction activities and the operation of these systems can lead to underwater noise, which may affect marine mammals and fish species. Engineers can play a crucial role in designing quieter systems and implementing mitigation measures to minimize the impact on these organisms.

Another significant consideration is the alteration of habitat and the potential for habitat loss. The installation of marine renewable energy devices, such as tidal turbines or wave energy converters, may require modifications to the seabed or coastline. It is essential for engineers to assess the potential impacts on benthic ecosystems, sediment dynamics, and coastal processes. By employing innovative engineering solutions, such as designing devices with minimal seabed disturbance or implementing habitat restoration programs, we can mitigate the adverse effects on marine habitats.

Furthermore, marine renewable energy systems can have visual and aesthetic impacts on coastal landscapes. Engineers can contribute to minimizing these impacts by integrating devices into the natural surroundings or implementing innovative design solutions that blend with the coastal environment.

The subchapter will also discuss the potential impacts of marine renewable energy systems on water quality, including the release of chemicals, nutrients, or thermal effluents. Engineers can develop robust monitoring systems and implement effective waste management strategies to prevent or mitigate any negative effects on water quality.

In conclusion, the environmental impact of marine renewable energy systems is a critical consideration for engineers in the field of geological engineering. By understanding and addressing these impacts through innovative design and engineering solutions, we can ensure that marine renewable energy systems contribute to a sustainable future without compromising the health of our oceans and marine ecosystems.

Monitoring and Assessment of Environmental Effects

Monitoring and assessment of environmental effects play a crucial role in the successful implementation of marine renewable energy systems. As engineers specializing in geological engineering, it is essential to understand the impact of these systems on the environment to ensure a sustainable future.

The subchapter "Monitoring and Assessment of Environmental Effects" aims to provide engineers with a comprehensive understanding of the methods and techniques used to monitor and assess the environmental effects of marine renewable energy systems.

One of the primary concerns associated with these systems is their potential impact on marine ecosystems. Engineers must be aware of the potential risks and be equipped with the knowledge to mitigate them. The subchapter will delve into the various monitoring techniques used to assess the effects on marine life, including the use of acoustic monitoring, underwater cameras, and tagging studies. It will also explore how engineers can collaborate with marine biologists and ecologists to gather relevant data and develop strategies for minimizing the impact on marine ecosystems.

Furthermore, the subchapter will discuss the monitoring and assessment of the physical and geological effects of marine renewable energy systems. Geological engineers play a crucial role in analyzing the potential impacts on coastal processes, sediment transport, and seabed stability. The content will address the methods used to monitor these effects, such as bathymetric surveys, geotechnical investigations, and numerical modeling. Engineers will learn how to assess the potential risks associated with changes in sediment transport, erosion, and deposition, and develop strategies to minimize these impacts.

Additionally, the subchapter will explore the monitoring and assessment of other environmental factors, such as noise emissions, electromagnetic fields, and visual impacts. Engineers will gain insights into the monitoring techniques used to assess these effects, including hydrophone arrays, electromagnetic field sensors, and visual impact assessments.

Overall, the subchapter on "Monitoring and Assessment of Environmental Effects" is designed to equip engineers specializing in geological engineering with the necessary knowledge and tools to evaluate and mitigate the potential environmental impacts of marine renewable energy systems. By understanding the monitoring techniques and assessment methods, engineers can contribute to the development of sustainable solutions for the future of marine renewable energy.

Mitigation Strategies for Environmental Impacts

As engineers in the field of geological engineering, it is crucial to consider the environmental impacts of marine renewable energy systems. While these systems offer sustainable solutions for our future energy needs, they can also pose potential risks to marine ecosystems. Therefore, it is essential to develop effective mitigation strategies to minimize these impacts and ensure the long-term sustainability of these projects.

One key mitigation strategy is conducting comprehensive environmental impact assessments (EIA) prior to the installation of marine renewable energy systems. EIAs help identify potential risks and impacts on marine life, water quality, and coastal habitats. By thoroughly assessing these impacts, engineers can design and implement appropriate measures to mitigate them.

To minimize the impact on marine life, engineers can consider utilizing technologies that reduce underwater noise emissions during the installation and operation of marine renewable energy systems. Underwater noise can adversely affect marine mammals, fish, and other marine organisms, disrupting their communication, migration, and feeding patterns. By incorporating noise-reducing technologies, such as bubble curtains or acoustic barriers, engineers can significantly mitigate these impacts.

Furthermore, engineers can explore the use of innovative monitoring systems to continuously assess the environmental impacts of marine renewable energy systems. These systems can measure parameters such as water quality, seabed stability, and marine biodiversity, providing real-time data for decision-making. By closely monitoring these parameters, engineers can quickly identify potential issues and take prompt action to mitigate any adverse effects.

In addition, engineers can collaborate with ecologists and biologists to develop and implement effective habitat restoration plans. These plans can include measures such as creating artificial reefs or establishing marine protected areas to enhance biodiversity and promote the recovery of ecosystems affected by the installation of marine renewable energy systems.

Educating the public and stakeholders about the environmental benefits and potential impacts of marine renewable energy systems is also crucial. By increasing awareness and promoting transparency, engineers can foster support and engagement from local communities, policymakers, and environmental organizations. This collaboration can lead to the development of more sustainable and environmentally friendly marine renewable energy projects.

In conclusion, as engineers in the field of geological engineering, it is our responsibility to develop and implement effective mitigation strategies for the environmental impacts of marine renewable energy systems. By conducting comprehensive environmental impact assessments, utilizing noise-reducing technologies, implementing innovative monitoring systems, and collaborating with ecologists and biologists, we can ensure the long-term sustainability of these projects while minimizing their adverse effects on marine ecosystems. Through education and collaboration, we can pave the way for a sustainable future powered by marine renewable energy.

Chapter 11: Economic and Policy Considerations

Cost Analysis of Marine Renewable Energy Systems

In the quest for sustainable energy solutions, marine renewable energy systems have emerged as promising alternatives to traditional fossil fuel-based power generation. As engineers specializing in geological engineering, it is crucial to understand the cost analysis associated with these innovative technologies. This subchapter aims to provide you with a comprehensive overview of the economic aspects of marine renewable energy systems.

Marine renewable energy systems encompass various technologies such as tidal, wave, and offshore wind energy. Each technology has its unique cost considerations, but they all share a common goal of harnessing the vast energy potential of the oceans. One of the key factors influencing the cost of these systems is the availability and accessibility of suitable marine sites. Geological engineers play a crucial role in identifying and characterizing these sites, ensuring optimal placement of marine renewable energy systems for maximum energy generation.

The cost analysis of marine renewable energy systems involves both capital and operational expenditures. Capital costs encompass the expenses associated with equipment procurement, installation, and grid connection. Geological engineers can contribute to cost reduction by leveraging their expertise in geological mapping and site characterization, helping to identify locations with favorable seabed conditions and minimizing foundation and anchoring costs.

Operational costs include ongoing maintenance, monitoring, and repair expenses. These costs are influenced by the system's reliability, accessibility, and distance from shore. By considering geological factors such as the presence of underwater geological formations or potential hazards, engineers can optimize maintenance strategies and reduce associated costs.

Furthermore, the subchapter delves into the concept of levelized cost of energy (LCOE), which provides a standardized metric to compare the cost competitiveness of different renewable energy technologies. LCOE takes into account the total lifecycle costs, including capital, operational, and decommissioning expenses, as well as the energy output over the system's lifetime. Geological engineers can contribute to optimizing LCOE by assessing and mitigating geological risks, ensuring the longevity and efficiency of marine renewable energy systems.

The subchapter also discusses the role of policy incentives and technological advancements in driving down costs and making marine renewable energy systems more economically viable. By highlighting successful case studies and cost reduction strategies, engineers specializing in geological engineering can enhance their understanding of the economic aspects of marine renewable energy systems.

In conclusion, the cost analysis of marine renewable energy systems is a crucial aspect of developing sustainable energy solutions. Geological engineers play a vital role in optimizing the costs associated with these technologies by identifying suitable marine sites, reducing capital and operational expenses, and considering factors such as geological risks and site accessibility. By understanding the economic aspects, engineers can contribute to the advancement and implementation of marine renewable energy systems, paving the way for a sustainable future.

Policy Frameworks and Regulatory Challenges

Policy Frameworks and Regulatory Challenges in Marine Renewable Energy Systems

As engineers specializing in geological engineering, it is vital to understand the policy frameworks and regulatory challenges surrounding marine renewable energy systems. Developing engineering solutions for a sustainable future requires a comprehensive knowledge of the legal and regulatory landscape to ensure the successful implementation of marine renewable energy projects.

The subchapter "Policy Frameworks and Regulatory Challenges" delves into the intricate web of policies and regulations that govern the deployment of marine renewable energy systems. It explores the various international, national, and regional policies that shape the industry and highlights the challenges faced by engineers working in this field.

At the international level, the subchapter discusses key initiatives such as the United Nations Convention on the Law of the Sea (UNCLOS) and the International Maritime Organization (IMO). These frameworks provide a legal basis for marine renewable energy activities and ensure the safety and sustainability of operations in international waters. Understanding these international policies is crucial for engineers engaged in projects that transcend national boundaries.

On the national and regional front, the subchapter examines the policies and regulations implemented by governments and authorities worldwide. It delves into the licensing processes, environmental impact assessments, and permitting requirements that engineers must navigate to obtain the necessary approvals for marine renewable energy projects. Additionally, it explores the financial incentives and support mechanisms offered by governments to promote the development of these systems.

The subchapter also highlights the unique challenges faced by engineers in the geological engineering niche. Geological considerations play a critical role in the design and implementation of marine renewable energy systems. Understanding the geological characteristics of the seabed and offshore environments is essential for site selection, foundation design, and cable routing. The subchapter discusses the regulatory aspects related to geological surveys, geotechnical investigations, and risk assessments that engineers must undertake to ensure the stability and integrity of these systems.

Furthermore, the subchapter addresses the evolving nature of policy frameworks and regulatory challenges. As technology advances and the industry matures, regulations need to adapt to new developments and address emerging concerns. The subchapter explores the need for flexible and adaptive policies that foster innovation while ensuring environmental protection and social acceptance.

In conclusion, "Policy Frameworks and Regulatory Challenges" is an essential subchapter for engineers specializing in geological engineering. It provides a comprehensive overview of the international, national, and regional policies that shape the marine renewable energy industry. By understanding and navigating these frameworks, engineers can contribute to the sustainable development of marine renewable energy systems, mitigating risks and ensuring compliance with regulations.

Financing and Market Trends in Marine Renewable Energy

As engineers in the field of geological engineering, it is important to stay updated on the latest financing and market trends in marine renewable energy. In recent years, the marine renewable energy sector has gained significant momentum due to the growing concerns over climate change and the need for sustainable energy sources. This subchapter aims to provide you with an overview of the financing options available and the market trends shaping the marine renewable energy industry.

One of the key challenges faced by marine renewable energy projects is securing adequate financing. Unlike conventional energy sources, marine renewable energy technologies are relatively new and require significant upfront investments. However, various financing options are available to support the development of these projects. Public funding, such as government grants and subsidies, play a crucial role in kick-starting marine renewable energy initiatives. Governments around the world are increasingly recognizing the importance of transitioning to clean energy sources and are allocating substantial funds to support research, development, and deployment of marine renewable energy systems.

In addition to public funding, private investments and partnerships have become vital in driving the growth of marine renewable energy. Venture capital firms, private equity funds, and institutional investors are increasingly interested in backing innovative marine renewable energy projects. This trend is largely driven by the potential for long-term returns on investment, as well as the increasing demand for sustainable energy solutions.

Furthermore, market trends in the marine renewable energy sector are also shaping the financing landscape. The industry is witnessing a shift towards commercial-scale projects, with a focus on reducing costs and improving efficiency. As a result, there is an increasing emphasis on project finance models, where financing is secured based on the project's revenue-generating potential. Power purchase agreements (PPAs) and feed-in tariffs (FITs) are commonly used mechanisms to ensure a stable revenue stream for marine renewable energy projects.

Moreover, market trends also reflect the growing interest in offshore wind energy. Offshore wind farms are becoming increasingly prevalent, with advancements in turbine technology and installation techniques. This presents new opportunities for geological engineers, as the design and installation of offshore wind turbines require expertise in marine geology and geotechnical engineering.

In conclusion, financing and market trends are crucial considerations for engineers in the field of geological engineering involved in marine renewable energy projects. Understanding the available financing options and staying updated on market trends will enable engineers to make informed decisions and contribute effectively to the development of sustainable marine renewable energy systems.

Chapter 12: Case Studies and Future Perspectives

Case Studies of Successful Marine Renewable Energy Projects

In recent years, the field of marine renewable energy has gained significant attention as the world seeks to transition towards more sustainable and clean energy sources. Engineers, especially those in the niche of geological engineering, have played a pivotal role in the development and implementation of various marine renewable energy projects. This subchapter will explore case studies of successful marine renewable energy projects, showcasing the ingenuity and engineering solutions employed to harness the power of the ocean.

One exemplary project that exemplifies the potential of marine renewable energy is the MeyGen tidal stream project in Scotland. This project, located in the Pentland Firth, has successfully deployed tidal turbines that harness the power of the tides to generate electricity. The engineering team faced numerous challenges, including extreme tidal conditions, complex seabed geology, and adverse weather conditions. Through meticulous planning and innovative design, the project has achieved remarkable success, generating clean and predictable energy for thousands of homes.

Another notable case study is the Cape Sharp Tidal project in the Bay of Fundy, Canada. Engineers in this project tackled the unique challenges of extremely high tidal ranges and strong currents. By developing specialized turbine designs and utilizing advanced geological survey techniques, the project has demonstrated the viability and potential of tidal energy in this region. The success of this project has paved the way for further exploration and development of marine renewable energy sources in the area.

Furthermore, the subchapter will discuss the EMEC (European Marine Energy Centre) project in Orkney, Scotland, which serves as a testbed for various marine renewable energy technologies. Engineers in this project have been instrumental in testing and evaluating novel devices, including wave energy converters and floating wind turbines. Their expertise in geological engineering has allowed for accurate site characterization and efficient resource assessment, enabling the successful integration of these technologies into the marine environment.

These case studies highlight the immense potential of marine renewable energy and the crucial role that engineers, particularly those in geological engineering, play in its development. By overcoming unique geological and environmental challenges, engineers have successfully harnessed the power of the ocean to generate clean and sustainable energy. These projects serve as inspirations and provide valuable lessons for future marine renewable energy initiatives, encouraging engineers to continue pushing the boundaries of what is possible in the pursuit of a sustainable future.

Emerging Technologies and Innovations

In the rapidly evolving field of marine renewable energy systems, engineers and scientists are constantly exploring new technologies and innovations to harness the power of the ocean in a sustainable and efficient manner. This subchapter delves into the exciting world of emerging technologies and innovations that are shaping the future of marine renewable energy, with a particular focus on their relevance to geological engineering.

One of the most promising emerging technologies is wave energy conversion systems. These systems capture the energy produced by ocean waves and convert it into electricity. Engineers are developing various types of wave energy converters, including point absorbers, oscillating water columns, and attenuators, each with its own advantages and challenges. Geological engineers can play a crucial role in the design and deployment of these systems by assessing the seabed conditions, ensuring stable foundations, and optimizing the placement of wave energy devices to minimize potential geological risks.

Another area of innovation is tidal energy. Tidal turbines, similar to wind turbines, are being developed to extract energy from tidal currents. These turbines are designed to withstand the harsh marine environment and provide a reliable and predictable source of renewable energy. Geological engineering expertise is vital in identifying suitable locations for tidal energy installations, considering factors such as sediment transport, seabed stability, and potential impacts on coastal erosion.

Floating offshore wind farms are also gaining momentum as a viable renewable energy solution. These innovative systems consist of wind turbines mounted on floating structures, allowing them to be deployed in deeper waters where traditional fixed-bottom wind farms are not feasible. Geological engineers can contribute by conducting geotechnical investigations to assess the seabed conditions and designing mooring systems that provide stability and support to these floating structures.

Furthermore, this subchapter explores the potential of marine current energy, ocean thermal energy conversion, and salinity gradient energy. Geological engineers can contribute to the development of these technologies by evaluating the geological and geotechnical aspects of potential project sites, ensuring the safe and reliable operation of marine renewable energy systems.

As the field of marine renewable energy continues to evolve, engineers specializing in geological engineering have a unique role in ensuring the successful integration of these innovative technologies. By considering geological factors in the design and implementation of marine renewable energy systems, engineers can contribute to a sustainable future by harnessing the power of the ocean while mitigating potential geological risks.

Future Trends and Opportunities in Marine Renewable Energy Systems

As engineers in the field of geological engineering, it is essential for us to stay updated with the latest trends and opportunities in marine renewable energy systems. The pursuit of sustainable energy sources has gained significant momentum in recent years, and marine renewable energy has emerged as a promising avenue for meeting our energy needs while minimizing environmental impacts.

One of the future trends in marine renewable energy systems is the advancement of tidal energy technology. Tidal energy harnesses the power of tidal currents to generate electricity. With the growing demand for clean energy, significant research and development efforts are being directed towards improving the efficiency and reliability of tidal energy systems. Engineers in the field of geological engineering can play a vital role in designing and optimizing these systems, considering the geological characteristics of the marine environment.

Another promising trend is the development of wave energy converters. Waves carry vast amounts of energy, and converting it into usable electricity can provide a significant contribution to our renewable energy mix. Engineers specializing in geological engineering can contribute by assessing the geological conditions and seabed characteristics to determine the most suitable locations for deploying wave energy converters. Furthermore, the development of innovative materials and engineering solutions to withstand the harsh marine environment is crucial for the successful implementation of wave energy systems.

Floating offshore wind farms are also gaining attention as a future trend in marine renewable energy systems. Traditional offshore wind farms require fixed foundations, which limit their deployment to shallow water areas. However, floating wind farms can be installed in deeper waters, opening up vast offshore regions for wind energy generation. Geological engineers can contribute by conducting detailed geotechnical investigations to ensure the stability and reliability of floating wind turbine structures in varying seabed conditions.

Moreover, the integration of marine renewable energy systems with other renewable energy sources and energy storage technologies presents exciting opportunities. Engineers specializing in geological engineering can contribute by assessing the geological and geotechnical aspects of energy storage solutions, such as compressed air energy storage in underground geological formations or pumped hydro storage in coastal areas.

In conclusion, the future of marine renewable energy systems holds immense potential for engineers in the field of geological engineering. By staying informed about the latest trends and opportunities in tidal energy, wave energy, floating offshore wind farms, and energy storage, we can actively contribute to the development of sustainable energy solutions for a greener future. As geological engineers, our expertise in assessing geological and geotechnical conditions will be invaluable in designing and optimizing marine renewable energy systems.

Chapter 13: Conclusion

Summary of Key Findings

In this subchapter, we will provide a comprehensive summary of the key findings presented in the book "Marine Renewable Energy Systems: Engineering Solutions for a Sustainable Future" specifically tailored for engineers in the niche of Geological Engineering. Throughout the book, the authors have explored various aspects of marine renewable energy systems, their engineering solutions, and their potential to contribute to a sustainable future.

One of the major findings highlighted in the book is the vast potential of marine renewable energy sources, such as tidal, wave, and offshore wind power. These sources can provide a significant amount of clean and renewable energy to meet the growing global demand. The authors emphasize the importance of understanding geological factors, such as seabed composition, geotechnical properties, and geological hazards, in the design and implementation of marine renewable energy systems. They highlight the need for engineers in the field of geological engineering to collaborate closely with other disciplines to ensure the successful development and operation of these systems.

The book also discusses the challenges associated with the installation and maintenance of marine renewable energy systems. It highlights the importance of site characterization and geotechnical investigations in selecting suitable locations for the deployment of these systems. The authors emphasize the need for engineers to consider geological factors, such as seabed stability, sediment transport, and erosion, to ensure the long-term viability and sustainability of these energy systems.

Furthermore, the book explores the potential environmental impacts of marine renewable energy systems and provides engineering solutions to mitigate these effects. It discusses the importance of understanding the interactions between marine renewable energy devices and the surrounding ecosystem, including marine life and habitats. The authors highlight the need for engineers to develop innovative technologies and design approaches that minimize the impact on marine environments while maximizing energy production.

In conclusion, "Marine Renewable Energy Systems: Engineering Solutions for a Sustainable Future" provides valuable insights for engineers in the field of Geological Engineering. It emphasizes the importance of considering geological factors in the design, deployment, and maintenance of marine renewable energy systems. The book highlights the significant potential of these systems to contribute to a sustainable future, while also discussing the challenges and environmental considerations associated with their implementation. By incorporating the findings presented in this book, engineers in the niche of Geological Engineering can play a crucial role in advancing marine renewable energy systems and ensuring a sustainable future for generations to come.

A Call for Collaboration and Sustainable Solutions in Marine Renewable Energy Systems

Introduction:

As engineers in the field of geological engineering, we are well aware of the numerous challenges posed by the ever-increasing energy demands of our society. Fossil fuel consumption has led to environmental degradation and climate change, necessitating a shift towards sustainable and renewable energy sources. In this subchapter, we address the urgent need for collaboration and sustainable solutions in marine renewable energy systems, offering insights into how geological engineers can contribute to this vital field.

The Potential of Marine Renewable Energy Systems:

Marine renewable energy systems, such as tidal, wave, and offshore wind energy, have immense potential to provide clean and renewable power. The vast energy reserves of our oceans remain largely untapped, and it is our responsibility as engineers to harness this potential. Geological engineers, with their expertise in studying the Earth's crust and its resources, can play a crucial role in the development of these systems.

Challenges and Collaborative Efforts:

Developing marine renewable energy systems presents unique challenges that demand interdisciplinary collaboration. Geological engineers can work alongside hydrologists, marine biologists, and environmental scientists to assess potential sites, understand the impact on marine ecosystems, and develop sustainable infrastructure. By integrating geological data and analysis into the design and implementation of these systems, we can ensure their long-term viability and minimize environmental impacts.

Innovative Engineering Solutions:

Geological engineers possess specialized knowledge in geotechnical analysis, seabed mapping, and offshore structure design, making them invaluable contributors to marine renewable energy projects. Through advanced geotechnical investigations, engineers can assess the stability of the seabed, design foundations for offshore turbines, and mitigate risks associated with geological hazards. Their expertise can also aid in the development of innovative materials and construction techniques that withstand the harsh marine environment.

The Role of Research and Development:

Research and development play a crucial role in advancing marine renewable energy systems. Geological engineers can contribute to this field by conducting research on resource assessment, environmental impact assessment, and optimizing the efficiency of energy conversion devices. By collaborating with academic institutions, industry experts, and government agencies, engineers can drive innovation and facilitate the transition towards a sustainable future.

Conclusion:

The urgency of addressing climate change and transitioning to renewable energy sources cannot be overstated. As geological engineers, we have a responsibility to collaborate and contribute our expertise to the development of marine renewable energy systems. By working together with other disciplines, we can design sustainable solutions that harness the power of the oceans while minimizing environmental impacts. Let us seize this opportunity to shape a sustainable future for generations to come.